Process Development Symposium 2023

Vancouver, Canada
13 - 15 June 2023

ISBN: 978-1-7138-8389-0

Printed from e-media with permission by:

Curran Associates, Inc.
57 Morehouse Lane
Red Hook, NY 12571

Some format issues inherent in the e-media version may also appear in this print version.

Printed with permission by Curran Associates, Inc. (2024)

AIChE
120 Wall Street, FL 23
New York, NY 10005-4020

Phone: (800) 242-4363
Fax: (203) 775-5177

www.aiche.org

Additional copies of this publication are available from:

Curran Associates, Inc.
57 Morehouse Lane
Red Hook, NY 12571 USA
Phone: 845-758-0400
Fax: 845-758-2633
Email: curran@proceedings.com
Web: www.proceedings.com

TABLE OF CONTENTS

POSTER SESSION

Chairmen's Introduction

Microwave industrial-scale processing found its earliest application in the food industry for drying, pasteurizing, and cooking. The motivations for using microwaves to replace conventional technologies related to order of magnitude reductions in processing time, improved product yields, and reduction of space required for processing. Today, the unique characteristics of microwave transfer of energy has led to its increasing use in many other industrial applications including ceramics processing, chemical synthesis, plasma processes including surface treatments, materials synthesis of both coatings and bulk materials, e.g., graphene and diamond, and more. We will review the benefits and limitations of microwave processing as a process intensification tool, present some specific examples of microwave processing hardware, and discuss potential future applications.

Growing Energy Supply While Lowering Carbon Intensity

Jablonski, C. - Presenter

Energy should be affordable, reliable and ever cleaner. Affordable energy is vital for economies to flourish. Reliable energy is essential for national security and could improve and even save lives. And we all have a stake in a lower carbon future. Sustainability for our energy future is dependent on growth in both traditional and new energy businesses because the worldâ€™s demand for energy is growing too. Process development and partnerships on a huge scale play vital roles in new energy solutions. A successful energy transition depends on all of us.

Scale-Down and Scale-up of a Slurry Mixing Process

Authors

Grenville, R. - Presenter, Philadelphia Mixing Solutions Ltd.
Giacomelli, J., Philadelphia Mixing Solutions Ltd.
Boyer, B., Philadelphia Mixing Solutions Ltd.

Scale-up is a common activity in process development. The goal is to develop the understanding of how a particular unit operation affects the outcome of the process in small scale equipment and then design equipment for the full-scale plant based on this understanding.

If a piece of equipment in a plant is not operating as designed, the operation can be scaled down, recreating the problem at small scale to develop an understanding of its causes. Then further tests can be carried out to develop a solution to the problem at small scale followed by the design of equipment to implement at plant scale.

A case study will be presented which will discuss how mixing issues were created, due to changes in the operating procedures, in the production of a viscous, non-Newtonian, high solids concentration slurry. The changes resulted in the slurry being more viscous and shear-thinning. The agitator was not capable of moving the slurry in regions far from the impeller blades and severe stagnation and fouling was observed. The process would have to be shut down on a regular basis so that the vessel could be cleaned.

Scale-down studies were carried out in a combined stirred tank / viscometer developed for the project enabling a mixing test to be performed identifying the degree of agitator required to eliminate stagnant regions and, at the same time, to measure the slurries rheology. The results were scaled-up and the new agitator has been working as specified for nearly five years.

Finally, some of the considerations that must be taken into account when making rheology measurements, especially for high concentration slurries, will be addressed.

Never Let a Good Crisis Go to Waste: RAPID RACER AMPD and the Opportunity to Accelerate Process Development – An Update

Authors

Gillenwater, P. - Presenter

As announced on March 7, 2022, the RAPID Manufacturing InstituteÂ® (an AIChE entity and one of the Manufacturing USA Institutes) was awarded $4.64MM by the U.S. Dept. of Commerceâ€™s National Institute of Standards and Technology (NIST) to both develop a methodology to accelerate process development and also develop a testbed for domestic manufacturing of critical pharmaceutical ingredients. The testbed will be located in the underserved Appalachian region in South Charleston, WV. The project is funded under the American Rescue Plan Act of 2021, also known as the COVID-19 Stimulus Package.

The test bed program, currently under development, presents an opportunity to accelerate chemical process development with modular skids and demonstrate its potential for rapid deployment of distributed manufacturing. Work is also underway to deliver an accelerated process development methodology to support the reshoring efforts that may take place during the next crisis/pandemic. We will present the learnings from our benchmarking survey of the AIChE Process Development Division on the current state of process development and how they are being codified into a stagewise methology for accelerated process development. Current work being done to explore the feasiblity of using a Damkohler number as a potential unifying tool for both the traditional reactor application as well as extension to downstream processing will also be discussed. An update on the test bed program will also be provided.

A Novel Zinc-Air battery Technology commercialization for Energy Storage application

Authors

Jauhar, M. A. - Presenter

Long-duration energy storage (LDES) technologies are advancing rapidly towards competitive deployment by enabling energy storage for extended demand periods with economic scalability. Zinc-air batteries have emerged as a promising candidate due to their high energy density, low cost, safety, and environmental friendliness. e-Zinc developed a breakthrough electrochemical technology for storing energy on-site through zinc-based battery systems. The system uses zinc metal as the energy carrier, providing high energy density at a lower cost. This solution offers flexible, long-duration energy storage for todayâ€™s low-carbon, dispatchable power needs and safety. This presentation discusses the topics of important factors in the scale-up and commercialization of the Zinc-air battery. We present different case studies to tackle scale-up challenges, such as optimizing of the re-circulation flow and synthesizing the electrolyte at the pilot scale.

Batch-to-Continuous Conversion of Substituted Alkyl Phenols

Authors

Al Azri, N. - Presenter
Cormack, G., Lubrizol
Proust, N.
Palchak, Z.
DeLetter, M.
Gasoline additives are a critical element of the global transportation fuels industry, specifically in North America where light passenger vehicle fuel requires mandatory additization. The global market for gasoline additives is circa 190,000MT/yr and is predominantly composed of three different chemistries: substituted alkylphenols, polyisobutylene (PIB) amines, and polyether amines. The premise of this project is to support market growth by minimizing capital and operating expenses of new capacity through modular chemical process intensification and batch-to-continuous conversion for production of substituted PIB phenol additives, known as Mannich detergents.

The current manufacturing process for the production of Mannich detergents involves the reaction of a PIB phenol with formaldehyde and a low molecular weight amine. The current batch process is executed in a 30MT reactor and entails 7-8 complex steps resulting in a 40-hour cycle time. The main driving forces for the long cycle time are mainly due to heat and mass transfer limitations associated with the high viscosity of the PIB at reaction temperature, the presence of a volatile amine above its vapor pressure, the presence of water as a by-product and diluent, and the need for exothermic reaction control. Currently, to resolve the viscosity limitations, a solvent is added which creates additional complexity during the post-reaction distillation of water and other volatiles.

In this work, the authors will present a case study on practical lessons learned through lab and pilot plant scale up of an intensified methodology for production of Mannich detergents. The study includes the minimization of heat and mass transfer limitations through novel unit operations, and the development of a kilo-scale continuous manufacturing unit. Finally, the authors will discuss practical considerations for modular unit scale up and deployment including OSBL infrastructure, raw materials, unit siting, controls integration, intermodal shipment, and financial analysis.

Chemical Transformers and the Clean Hydrogen Grid

Authors

Wegeng, R. - Presenter, STARS Technology Corporation

Compact Hydrogen Generators, placed as â€œchemical transformersâ€• on the natural gas grid, can be compared to electrical transformers on the electrical grid. Reconditioning the natural gas (or renewable natural gas) supply to produce hydrogen at or near points-of-use, is a near-term opportunity to begin converting the existing gas grid infrastructure into a highly functional, clean hydrogen grid.

Of critical interest:

- This approach avoids the high cost of transporting hydrogen from central facilities to points of use, a problem that has substantially constrained the rollout of hydrogen filling stations and fuel cell electric vehicles in California, and
- When renewable natural gas is used as the feedstock, the hydrogen product will be scored as low-carbon or carbon-negative using GREET-type assessments.

This presentation will describe a demonstration project that is currently underway in southern California, showing how compact process-intensive microchannel reactors and heat exchangers enable the deployment of mass-producible, low-cost â€œchemical transformersâ€• as hydrogen generators. Deployed in this manner, the chemical transformers enable low-carbon and even carbon-negative hydrogen at affordable costs for transportation and other uses.

Process-intensive hardware enables high-efficient, modular chemical transformers with small footprints and volumes that can be mass-produced at low cost using assembly lines. We project that the levelized production cost of the hydrogen at the point of use, using these systems, will be less than \$3/kg, which based on low fossil carbon emissions would be eligible for subsidies â€“ such as the Clean Hydrogen Production Tax Credit â€“ that further reduces costs to hydrogen users.

This technology demonstration project strongly supports the ambition and need to rapidly reduce fossil carbon emissions in order to limit, and then halt, the rise in atmospheric greenhouse gas levels. In the Pacific Northwest, transportation emissions are more than half of all anthropogenic emissions. In order to support the goal of reducing emissions by 50% by 2030, and to support the goals of the 2015 Paris Agreement, we must aggressively establish affordable, clean H_2 infrastructure promptly. These outcomes are achievable through the deployment of Compact Hydrogen Generators as â€œchemical transformersâ€• on the gas grid, making affordable, clean H_2 available throughout North America and providing a template for others throughout the world.

Participants in the demonstration project, along with STARS Technology Corporation, include Southern California Gas Company, SunLine Transit Agency, the US Department of Energyâ€™s Pacific Northwest National Laboratory and RAPID Institute, and several other companies.

Microwave Assisted Processing for Industrial Scale Material and Chemical Synthesis Applications

Authors

Weiss, R. - Presenter

Microwave industrial-scale processing found its earliest application in the food industry for drying, pasteurizing, and cooking. The motivations for using microwaves to replace conventional technologies related to order of magnitude reductions in processing time, improved product yields, and reduction of space required for processing. Today, the unique characteristics of microwave transfer of energy has led to its increasing use in many other industrial applications including ceramics processing, chemical synthesis, plasma processes including surface treatments, materials synthesis of both coatings and bulk materials, e.g., graphene and diamond, and more. We will review the benefits and limitations of microwave processing as a process intensification tool, present some specific examples of microwave processing hardware, and discuss potential future applications.

Assemblytron: Flexible Automation of DNA Assembly with Opentrons OT-2 Lab Robots

Authors

Bryant, J. Jr. - Presenter

As one of the newest fields of engineering, synthetic biology relies upon a trial-and-error Designâ€“Buildâ€“Testâ€“Learn (DBTL) approach to simultaneously learn how a function is encoded in biology and attempt to engineer it. Many software and hardware platforms have been developed to automate, optimize and algorithmically perform each step of the DBTL cycle. However, there are many fewer options for automating the build step. Build typically involves deoxyribonucleic acid (DNA) assembly, which remains manual, low throughput and unreliable in most cases and limits our ability to advance the science and engineering of biology. Here, we present AssemblyTron, an open-source Python package to integrate j5 DNA assembly design software outputs with build implementation in Opentrons liquid handling robotics with minimal human intervention. We demonstrate the versatility of AssemblyTron through several scarless, multipart DNA assemblies, beginning from fragment amplification. We show that AssemblyTron can perform polymerase chain reactions across a range of fragment lengths and annealing temperatures by using an optimal annealing temperature gradient calculation algorithm. We then demonstrate that AssemblyTron can perform Golden Gate and homology-dependent *in vivo* assemblies (IVAs) with comparable fidelity to manual assemblies by simultaneously building four four-fragment assemblies of chromoprotein reporter expression plasmids. Finally, we used AssemblyTron to perform site-directed mutagenesis reactions via homology-dependent IVA also achieving comparable fidelity to manual assemblies as assessed by sequencing. AssemblyTron can reduce the time, training, costs and wastes associated with synthetic biology, which, along with open-source and affordable automation, will further foster the accessibility of synthetic biology and accelerate biological research and engineering.

Opportunities for Process Development Engineers in Accelerating the Transport Sector Energy Transition

Authors

Bej, S. - Presenter, Shell
Opportunities for Process Development Engineers in Accelerating the Transport Sector Energy Transition

Shyamal Bej

Abstract:

The energy mixture presently used to fuel the transport sector is predominantly of fossil origin and leads to significant net carbon dioxide emissions. As a result, there is an urgent need to transition the existing energy mixture into one which consists of renewable sources and has minimal fossil origin. After giving a brief overview of the history of previous energy transitions and the vast amount of knowledge generated while working through these transitions, this talk will focus on future fuels proposed for the transport sector and the technologies required to produce these future fuels. A summary of examples where knowledge gained during earlier transitions can be leveraged and where new technologies need to be developed will be reviewed. Enormous opportunities exist for process development engineers in improving current technologies as well as developing new technologies for successfully achieving the goals of the current energy transition.

Sustainability in Modular Process Plants

Authors

Edwards, D. - Presenter, Zeton Inc
Sustainability in the CPI typically focuses on making the process chemistry more benign, or using renewable sources of raw materials, or how products can be designed with recycling in mind. But what about the plant itself? What is being done to make the design, fabrication, installation, operation and eventual decommissioning of process plants more sustainable? This article will focus on modular process plants, and what work is underway to reduce the environmental footprint in their design and fabrication, offering a comparison to the traditional stick-built approach.

Utilizing Azeotropes & Solvent Recycle for Cost Effective Implementation on Batch 1

Authors

Couling, D. - Presenter, Massachusetts Institute of Technology

This talk describes the development and scale-up of a solvent recovery strategy that utilized fundamental experimental data to support successful commercial implementation. The commercialization of an agrochemical intermediate required solvent recovery within a triethylamine-acetonitrile-water solvent system. The design of the recovery process was inhibited by the lack of reliable experimental data for this mixture. A spinning band column was used to determine the equilibrium azeotropic compositions. A successful process was developed based on these data to maximize the recovery of all three species. Additionally, the implementation of the solvent recovery improved the downstream crystallization and reduced the amount of waste generated. These detailed studies proved invaluable during start-up and were required on the very first batch due to an unforeseen global shortage of acetonitrile.

Thermochemical Production of Green Hydrogen or Syngas

Authors

Boughey, B. - Presenter
Warren, K. J., University of Colorado Boulder

Meeting growing domestic energy demands while reducing our nationâ€™s carbon footprint requires the development of technologies that not only harness the abundant renewable resources of the United States (e.g., concentrated solar radiation and renewable electricity) but also allow for such energy to be dispatched on demand to the end user, regardless of the time of day or geographic location. One particularly attractive approach is the production of syngas via thermochemical water and carbon dioxide splitting, as such pathway â€“ unlike alternative electrolysis-based technologies â€“ relies entirely on the conversion of process heat which better complements traditional economies of scale. Importantly, syngas, when integrated with well-established catalytic processes like Fischer-Tropsch synthesis, can be converted into stable, transportable hydrocarbons that are compatible with existing infrastructure.

In this talk, we will present an innovative thermochemical process using a novel metal oxide that enables breakthrough energy efficiencies in the production of syngas or green hydrogen. This process operates isothermally, eliminating more than 60% of the losses faced by current state of the art thermochemical water or carbon dioxide splitting technologies that incorporate wide temperature swings. If operating solely from electrical heat, even higher efficiencies can be realized as re-radiative losses associated with concentrated solar thermal heat sources would be eliminated. The opportunity to provide process heat via low-cost and low-carbon sources of electricity also unlocks the potential for a reactor system that can be co-sited with other industrial processes, potentially scavenging waste heat to comprise some of the process heat requirements and thus further lowering OPEX.

Utilizing Cross-functional Collaboration for the Successful Implementation of Commercial Scale-up

Authors

Marple, D. - Presenter, Dow Performance Silicones

In industry, the commercialization of many new products begins with studies and data collections conducted on laboratory or benchtop equipment. In order to be produced commercially, a scale up to full-scale assets is required. Due to a wide variety of materials and processes, there is no single scale-up method that can be applied, and several approaches exist.

This poster looks at a case study of scaling up a new material in which a cross-functional team was utilized to help in troubleshooting various issues that arose during the process. Ultimately the scale-up was successful and the material is being produced commercially.

Demonstration of a Process Intensification Scenario for Producing Ammonia Synthesis Gas without an Air Separation Unit Using Soec Technology

Authors

Helmig, K. - Presenter
Partoon, B., Universiti Teknologi Petronas
Dannesboe, C.
Hansen, J. B., Haldor Topsøe A/S
Oestberg, M., Haldor Topsoe A/S
Like conventional technologies, green ammonia production usually produces nitrogen through a cryogenic Air Separation Unit (ASU). An ASU requires up to 10% of CAPEX and 1% of electricity in a green ammonia plant. With recent development in Solid Oxide Electrolyzer Cells (SOEC) technologies, by proper process intensification, ASU could be eliminated from the green ammonia plant. A SOEC unit produces hydrogen from steam at high temperatures (700-800 °C), opening up heat integration potential with the exothermic Haber Bosch reaction. As a bonus, the process utilize the unique capabilities of SOEC to act as an oxygen separation membrane and its ability to utilize heat in lieu of power. The burning of air generates extra steam, which further could be electrolyzed again. Still, due to the high efficiency of SOEC, being close to 100 %, the net result is only a marginal increase in overall power consumption. The process ensures far better scalability of an electrified ammonia plant, as a much lower investment can be realized due to the elimination of the air separation unit. This work details this new synthesis gas production process and presents experimental results demonstrating 1000 hours of operation of this novel production pathway using a 25 kW synthesis gas production unit (10 Nm3/h ammonia synthesis gas).

AUTHOR INDEX